A NATIONAL STRATEGIC PLAN FOR ADVANCED MANUFACTURING

Executive Office of the President

National Science and Technology Council

FEBRUARY 2012

A NATIONAL STRATEGIC PLAN FOR ADVANCED MANUFACTURING

Executive Office of the President

National Science and Technology Council

FEBRUARY 2012

About the National Science and Technology Council

The National Science and Technology Council (NSTC) is the principal means by which the Executive Branch coordinates science and technology policy across the diverse entities that make up the Federal research and development enterprise. A primary objective of the NSTC is establishing clear national goals for Federal science and technology investments. The NSTC prepares research and development strategies that are coordinated across Federal agencies to form investment packages aimed at accomplishing multiple national goals. The work of the NSTC is organized under five committees: Environment, Natural Resources and Sustainability; Homeland and National Security; Science, Technology, Engineering, and Math (STEM) Education; Science; and Technology. Each of these committees oversees subcommittees and working groups focused on different aspects of science and technology. More information is available at http://www.whitehouse.gov/ostp/nstc.

About the Office of Science and Technology Policy

The Office of Science and Technology Policy (OSTP) was established by the National Science and Technology Policy, Organization, and Priorities Act of 1976. OSTP's responsibilities include advising the President in policy formulation and budget development on questions in which science and technology are important elements; articulating the President's science and technology policy and programs; and fostering strong partnerships among Federal, state, and local governments, and the scientific communities in industry and academia. The Director of OSTP also serves as Assistant to the President for Science and Technology and manages the NSTC. More information is available at http://www.whitehouse.gov/ostp.

About the Interagency working group on Advanced Manufacturing

The Interagency working group on Advanced Manufacturing (IAM) serves as part of the internal deliberative process of the NSTC and provides overall guidance and direction on advanced manufacturing matters. The IAM serves as a forum within the NSTC for developing consensus and resolving issues associated with advanced manufacturing policy, programs, and budget guidance. The goals of the IAM are to (a) identify and integrate technical requirements, (b) conduct joint program planning and coordination, and (c) develop joint strategies or multi-agency joint solicitations for advanced manufacturing programs conducted by the Federal government. The IAM serves as a forum for the exchange and leveraging of information among the participating agencies.

About this Document

This report was developed by the Interagency working group on Advanced Manufacturing (IAM). The IAM reports to the NSTC Committee on Technology. This report is published by the Office of Science and Technology Policy

Copyright Information

This document is a work of the United States Government and is in the public domain (see 17 U.S.C. §105). Subject to the stipulations below, it may be distributed and copied with acknowledgment to OSTP. Copyrights to graphics included in this document are reserved by the original copyright holders or their assignees and are used here under the government's license and by permission. Requests to use any images must be made to the provider identified in the image credits or to OSTP if no provider is identified.

Printed in the United States of America, 2012.

Report prepared by
NATIONAL SCIENCE AND TECHNOLOGY COUNCIL
COMMITTEE ON TECHNOLOGY (CoT)
INTERAGENCY WORKING GROUP ON ADVANCED MANUFACTURING (IAM)

National Science and Technology Council

Chair

John P. Holdren
Assistant to the President for
Science and Technology
Director
Office of Science and Technology Policy

Staff

Pedro I. Espina
Executive Director

Committee on Technology

Chair

Tom Power*
Deputy Chief Technology Officer
of the United States
Office of Science and Technology Policy

Staff

Pedro I. Espina
Executive Secretary

Interagency working group on Advanced Manufacturing

Co-chairs

Gregory Tassey
Chief Economist
National Institute of Standards and Technology
Department of Commerce

Adele Ratcliff
Director, Manufacturing Technology
Office of the Secretary of Defense
Department of Defense

Leo Christodoulou
Program Manager, Advanced Manufacturing
Energy Efficiency and Renewable Energy
Department of Energy

Staff

Robert W. Ivester
Executive Secretary

Members

Daniel Schmoldt
National Program Leader
Institute of Food Production and Sustainability
National Institute of Food and Agriculture
Department of Agriculture

Nathan Wiedenman
Program Manager, Tactical Technology Office,
Defense Advanced Research Projects Agency
Department of Defense

---

* Formerly Aneesh Chopra

Frank Chong
Office of Vocational and Adult Education
Department of Education

Gregory Henschel
Office of Vocational and Adult Education
Department of Education

Thomas Cellucci
Chief Commercialization Officer
Science and Technology Directorate
Department of Homeland Security

Sean Cartwright
Chief of Staff
Employment and Training Administration
Department of Labor

Amy Young
Senior Advisor to the Assistant Secretary
Employment and Training Administration
Department of Labor

Curtis Tompkins
Acting Associate Administrator for Research, Development and Technology, Research and Innovative Technology Administration
Department of Transportation

Alan Hecht
Director for Sustainable Development
Office of Research and Development
Environmental Protection Agency

Rosemarie Hunziker
Program Director, National Institute of Biomedical Imaging and Bioengineering
National Institutes of Health

Steven McKnight
Director, Civil, Mechanical, and Manufacturing Innovation Division
National Science Foundation

Bruce Kramer
Senior Advisor, Civil, Mechanical, and Manufacturing Innovation Division
National Science Foundation

Grace Hu
Program Examiner
Office of Management and Budget

David Hart
Associate Director for Innovation Policy
Office of Science and Technology Policy

Sridhar Kota
Assistant Director for Advanced Manufacturing
Office of Science and Technology Policy

Acknowledgements

A number of key contributors provided significant support to the activities of the working group and the development of this report.

Department of Defense
R. Scott Frost (ANSER), Mark Gordon (NCAT), Brynne Ward (OSD)

Department of Energy
Jamie Link (EERE)

Department of Commerce
Albert Jones (NIST), Gary Yakimov (NIST), Phillip Singerman (NIST), Bessmarie Young (NIST)

Department of Homeland Security
Mark Protacio (S&T)

Environmental Protection Agency
Meadow Anderson (AAAS Fellow for Science and Technology)

EXECUTIVE OFFICE OF THE PRESIDENT
NATIONAL SCIENCE AND TECHNOLOGY COUNCIL
WASHINGTON, D.C. 20502

February 15, 2012

Members of Congress:

I am pleased to transmit with this letter the report of the National Science and Technology Council's Interagency working group on Advanced Manufacturing (IAM). This report responds to Section 102 of the America COMPETES Reauthorization Act of 2010, which directs the NSTC to develop a strategic plan to guide Federal programs and activities in support of advanced-manufacturing research and development. It builds on *The Report to the President on Ensuring American Leadership in Advanced Manufacturing*, which was released by the President's Council of Advisors on Science and Technology (PCAST) in June 2011.

Advanced manufacturing is a matter of fundamental importance to the economic strength and national security of the United States. The analysis of patterns and trends in U.S. advanced manufacturing contained in this NSTC report reveals both opportunities for Federal policy to accelerate the development of this vital sector and challenges to its continuing health. The NSTC's work, together with PCAST's recommendations and the forthcoming report of the Advanced Manufacturing Partnership Steering Committee, provides a solid foundation for a Federal policy that will enable the United States to "build it here, and sell it everywhere," as Secretary of Commerce John Bryson put it recently.

I look forward to working with the Congress and other key partners to realize that goal.

Sincerely,

John P. Holdren
Assistant to the President for Science and Technology
Director, Office of Science and Technology Policy

Table of Contents

Executive Summary 1

1. Introduction 2

2. Advanced Manufacturing: Patterns and Trends 4
    Global Trends in Advanced Manufacturing 4

3. Principles and Objectives of the National Strategy 7
    Innovation Policy for Advanced Manufacturing 7
    Strengthening the Industrial Commons 8
    Optimizing Federal Investments 9
    Stakeholders and Objectives 10

4. Accelerating Investment by Small and Medium-Sized Enterprises 11
    Private-Public Co-Investment 11
    Early Procurement 12
    Advanced Manufacturing for National Security 12
    Metrics and Key Implementing Agencies for Objective 1 13

5. Strengthening Workforce Skills 14
    The Changing Manufacturing Workforce 15
    Better Training for Today's Advanced Manufacturing Workers 15
    Education and Training for Tomorrow's Workers 16
    Educating the Next Generation 17
    Metrics and Key Implementing Agencies for Objective 2 18

6. Creating Partnerships 19
    SME Engagement through Partnerships 19
    Cluster-Based Partnerships 20
    Metrics and Key Implementing Agencies for Objective 3 21

7. Coordinating Federal Investments 22
    The Federal Advanced Manufacturing Investment Portfolio 22
    Advanced Materials 23
    Product Technology Platforms 23
    Advanced Manufacturing Processes 24
    Data and Design Infrastructure 24
    Cross-Cutting Agency Investments 25
    Metrics and Key Implementing Agencies for Objective 4 25

8. Raising National Investment in Advanced Manufacturing R&D 26
    R&E Tax Credit. 26
    Federal Investment 27
    Metrics and Key Implementing Agencies for Objective 4.. 27

Appendix A: America COMPETES Reauthorization Act of 2010 Section 102 28

Appendix B: The Defense Production Act Committee (DPAC) 30

Appendix C: The Department of Defense Manufacturing Technology Program 31

Appendix D: Examples of Federal Support for the Advanced Manufacturing Workforce 33

Appendix E: The Advanced Manufacturing Competency Model of the Department of Labor . . 35

Appendix F: Examples of Federal Investments in Advanced Manufacturing R&D 36

Executive Summary

This report responds to Section 102 of the America COMPETES Reauthorization Act of 2010, which directs the Committee on Technology of the National Science and Technology Council (NSTC) to develop a strategic plan to guide Federal programs and activities in support of advanced manufacturing research and development. Advanced manufacturing is a matter of fundamental importance to the economic strength and national security of the United States. Our analysis of patterns and trends in U.S. advanced manufacturing reveals both opportunities for Federal policy to accelerate the development of this vital sector and challenges to its continuing health.

The acceleration of innovation for advanced manufacturing requires bridging a number of gaps in the present U.S. innovation system, particularly the gap between research and development (R&D) activities and the deployment of technological innovations in domestic production of goods. This strategic plan lays out a robust innovation policy that would help to close these gaps and address the full lifecycle of technology. It also incorporates intensive engagement among industry, labor, academia, and government at the national, state, and regional levels. Partnerships among diverse stakeholders, varying by location and objective, are a keystone of the strategy.

The strategy seeks to achieve five objectives. These objectives are interconnected; progress on any one will make progress on the others easier. A large number of Federal agencies, coordinated through the NSTC, have important roles to play in the implementation of the strategy.

Objective 1: *Accelerate investment in advanced manufacturing technology, especially by small and medium-sized manufacturing enterprises, by fostering more effective use of Federal capabilities and facilities, including early procurement by Federal agencies of cutting-edge products.*

Objective 2: *Expand the number of workers who have the skills needed by a growing advanced manufacturing sector and make the education and training system more responsive to the demand for skills.*

Objective 3: *Create and support national and regional public-private, government-industry-academic partnerships to accelerate investment in and deployment of advanced manufacturing technologies.*

Objective 4: *Optimize the Federal government's advanced manufacturing investment by taking a portfolio perspective across agencies and adjusting accordingly.*

Objective 5: *Increase total U.S. public and private investments in advanced manufacturing research and development (R&D).*

1. Introduction

Section 102 of the America COMPETES Reauthorization Act of 2010 directs the Committee on Technology of the National Science and Technology Council (NSTC) to develop a strategic plan to guide Federal programs and activities in support of advanced manufacturing research and development (see Appendix A). This report responds to that provision. It was prepared by the Interagency Working Group on Advanced Manufacturing (IAM) of the NSTC, which included representatives of Federal agencies with a vital interest in sustaining and strengthening the Nation's advanced manufacturing sector.

Advanced manufacturing is a family of activities that (a) depend on the use and coordination of information, automation, computation, software, sensing, and networking, and/or (b) make use of cutting edge materials and emerging capabilities enabled by the physical and biological sciences, for example nanotechnology, chemistry, and biology. It involves both new ways to manufacture existing products, and the manufacture of new products emerging from new advanced technologies.[1]

Advanced manufacturing is a matter of fundamental importance to the economic strength and national security of the United States. Advanced manufacturing provides high-quality jobs. It is an important source of exports. It is a key source of technological innovation. It provides essential goods and equipment for the military, the intelligence community, and homeland security agencies. These impacts justify Congressional and executive branch attention to Federal policies that affect advanced manufacturing.

Our analysis of patterns and trends in U.S. advanced manufacturing reveals both opportunities for Federal policy to accelerate the development of this vital sector and challenges to its continuing health. In particular, a gap exists between research and development (R&D) activities and the deployment of technological innovations in domestic production of goods.[2] This gap has contributed to the erosion of key indicators, such as the balance of trade in advanced technology products as measured by the U.S. Census Bureau (see Figure 1).[3] The United States ran a trade surplus in this category throughout the 1990s, but by 2010, that surplus had become an $81 billion deficit.

---

1. President's Council of Advisors on Science and Technology, *Report to the President on Ensuring American Leadership in Advanced Manufacturing*, June 2011, p. ii.
2. This report does not address broader issues such as Federal tax (with the exception of the research and experimentation tax credit), infrastructure, investment, intellectual property, trade, and export-promotion policies that are associated with competitiveness and innovation but that are explicitly assigned in Section 604 of COMPETES to a separate report from the Department of Commerce, which was published in January 2012. See http://www.commerce.gov/americacompetes.
3. The Census Bureau defines Advanced Technology Products using about 500 of some 22,000 commodity classification codes used in reporting U.S. merchandise trade. Each of the 500 codes meets the following three criteria – (1) the code contains products whose technology is from a recognized high technology field, (2) these products represent leading edge technology in that field, and (3) such products constitute a significant part of all items covered in the selected classification code.

1. INTRODUCTION

This strategic plan lays out a robust innovation policy that would reduce the gap between R&D and deployment of advanced manufacturing innovations. Specifically, this policy would address the full lifecycle of technology in order to (1) provide a fertile innovation environment for advanced manufacturing, (2) enable vigorous domestic development of transformative manufacturing technologies, (3) promote coordinated public and private investment in precompetitive advanced manufacturing technology infrastructure, and (4) facilitate rapid scale-up and market penetration of advanced manufacturing technologies.[4]

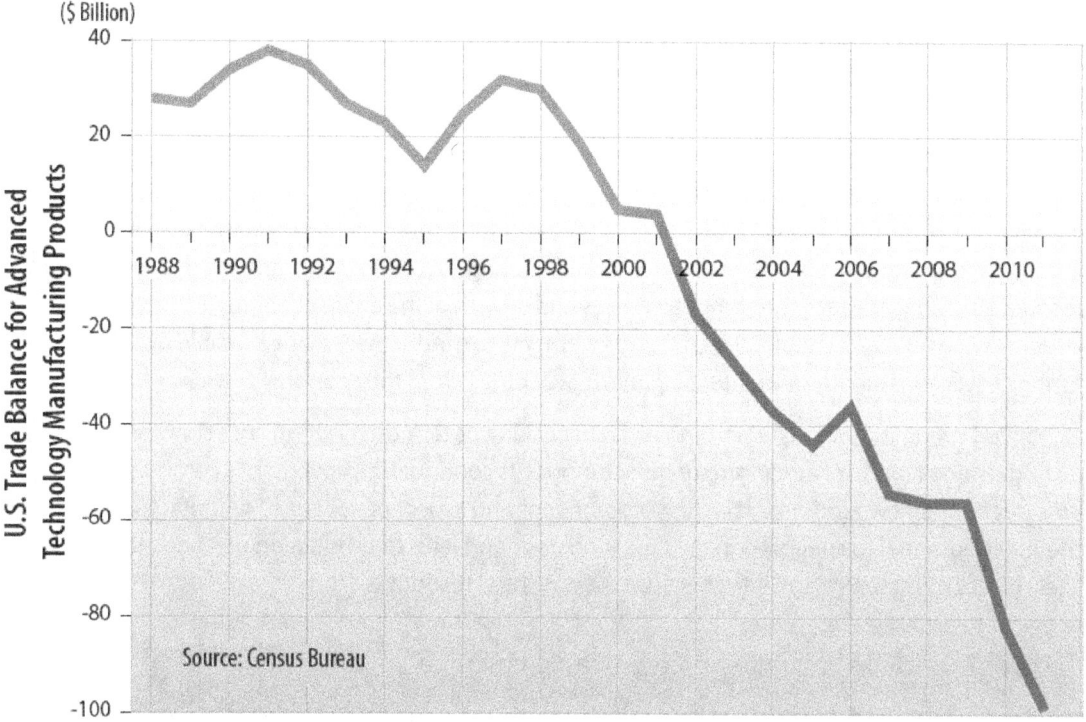

Figure 1. U.S. Trade Balance for Advanced Technology Products

Federal policies and programs alone cannot address the Nation's challenges in advanced manufacturing. The strategic plan therefore incorporates intensive engagement among industrial, labor, academic, and government stakeholders at the national, state and regional levels.

The next section describes the challenges and opportunities for advanced manufacturing in more detail. Section 3 articulates the broad approach of the strategic plan and its five objectives. The remaining sections elaborate on how each of these objectives might be achieved.

---

4. PCAST, *Advanced Manufacturing*.

2. Advanced Manufacturing: Patterns and Trends

U.S. manufacturers produced about $1.7 trillion of goods in 2010, about 11.7% of the U.S. gross domestic product (GDP).[5] They employed 11.5 million Americans in jobs that paid on average about 21% more than average hourly compensation in private-sector service industries.[6] Manufacturing has a larger multiplier effect than any other major economic activity -- a dollar spent in manufacturing drives an additional $1.35 in economic activity.[7] Manufacturing is also the largest contributor to U.S. exports. In 2010, the United States exported over $1.1 trillion of manufactured goods, which accounted for 86% of all U.S. goods exports and 60% of U.S. total exports.[8]

Manufacturing provides many of the jobs and drives many of the businesses of today. Yet its role in providing the jobs and driving the businesses of tomorrow is even more important. The manufacturing sector accounts for about 72% of all private-sector R&D spending and employs about 60% of U.S. industry's R&D workforce.[9] As a result, the manufacturing sector develops and produces many of the technologies that advance the competitiveness and growth of the entire economy, including the much larger service sector. Technology-based improvements to productivity made possible by the manufacturing sector consistently generate job growth over time across the economy.[10]

Advanced manufacturing is emerging as an especially potent driver of future economic growth. A distinguishing feature of advanced manufacturing is its continual improvement in processes and rapid introduction of new products. It is this paradigm-shifting aspect of advanced manufacturing that has the most potential to spin off entirely new industries and lead to production methods that are most likely to "stick" in the United States because they are hard to imitate.

Global Trends in Advanced Manufacturing

Current global trends in R&D, innovation, and trade raise concerns about U.S. competitiveness in advanced manufacturing. In 2009, the United States ranked eighth among industrialized nations for R&D intensity (defined as national R&D as a share of GDP), according to the Organisation for Economic Co-operation and Development (OECD).[11] A 2011 report by the Information Technology and Innovation Foundation ranked the United States fourth out of 44 industrialized countries and regions in global innovative-based competitiveness, but second-to-last in progress toward increasing innovation-based competitiveness and capacity since 2000.[12]

---

5. Bureau of Economic Analysis, 2010 U.S. Economic Accounts by Industry, see http://www.bea.gov/industry/index.htm.
6. Bureau of Labor Statistics, 2011 *Employer Costs for Employee Compensation*, Table 6.
7. Bureau of Economic Analysis, *Industry-by-Industry Total Requirements Table*, see http://www.bea.gov/industry/iotables/prod/.
8. Bureau of Economic Analysis and Census, *U.S. International Trade in Goods and Services*.
9. National Science Board, *Science and Engineering Indicators* 2012, Appendix Table 4-14 and Table 3-32.
10. Peter Bisson, Elizabeth Stephenson, and S. Patrick Viguerie, "The Productivity Imperative," *McKinsey Quarterly*, June 2010.
11. National Science Foundation, *Science and Engineering Indicators* 2012, p. 4-42, see http://www.nsf.gov/statistics/seind12/.
12. R. Atkinson and S. Andes, *The Atlantic Century II: Benchmarking E.U. and U.S. Innovation and Competitiveness*. Washington, DC: Information Technology and Innovation Foundation, 2011.

2. ADVANCED MANUFACTURING: PATTERNS AND TRENDS

As we noted in the Introduction, the Nation's trade balance for advanced technology products has deteriorated precipitously over the past decade, despite an offsetting 34% decline in the major-currency foreign exchange value of the U.S. dollar.[13] Currently, Germany, Korea, and Japan each have more R&D-intensive manufacturing sectors than the United States (see Figure 2);[14] moreover, they each have positive trade balances in goods.

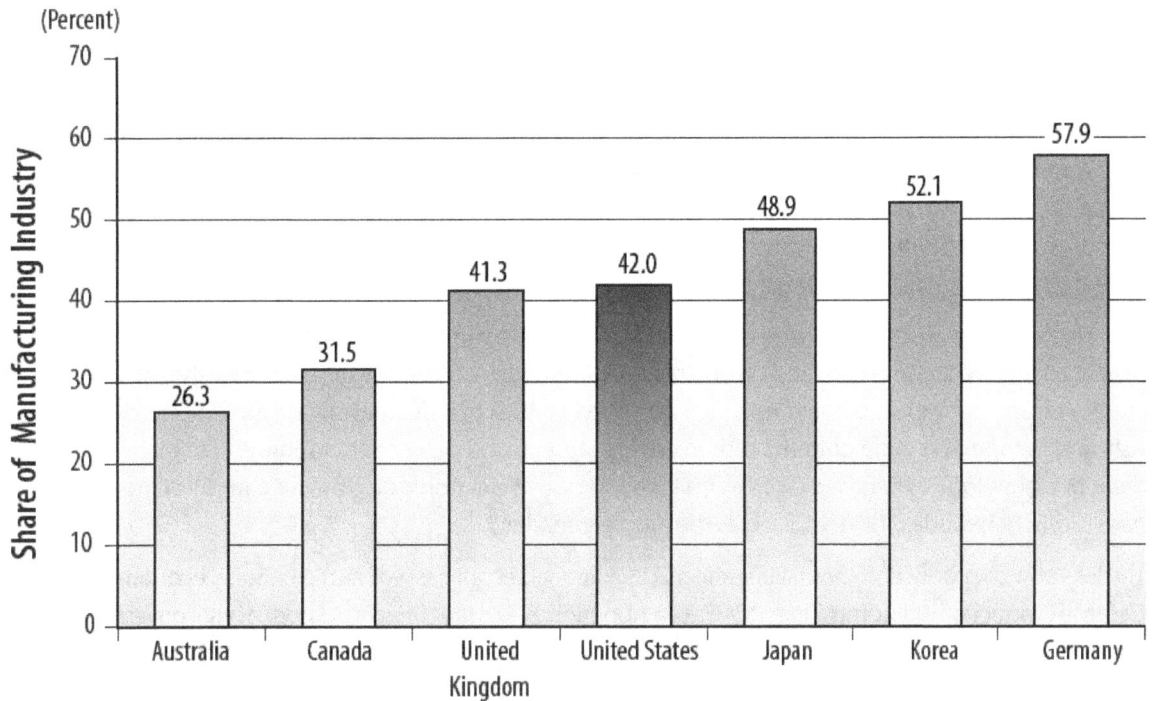

Source: OECD STAN Indicators 2009, "Value-added shares relative to manufacturing, " Percent of manufacturing sector with 3% or greater R&D intensity, stats.oecd.org/index.aspx?r=228903

Figure 2. Share of Manufacturing Value-Added by Research-Intensive Manufacturing Sectors (R&D > 3% of Sales)

---

13. The U.S. "major-currency" dollar index is the value (weighted geometric mean) of the dollar relative to a basket of foreign currencies (Euro, English pound, Canadian dollar, Swedish krona, Swiss franc, and Japanese yen). During this same period, the dollar declined 22 percent against an index of all currencies. See http://www.Federalreserve.gov/releases/h10/summary/default.htm.
14. R&D intensity is measured by R&D as a share of industry sales; R&D-intensive industries have R&D/sales ratios of greater than 3%.

Global competition in advanced manufacturing is growing more intense as technology lifecycles are accelerated. A lifecycle starts with the basic concept for a technology. For many technologies, scientific knowledge created through basic research provides key insights that enable the basic concept. Then the concept is validated, often through applied research. Next, development efforts mature the technology into a prototype of a commercial product. Finally, commercialization and scale-up activities convert the prototype into a commercially viable product and increase the scale of production to an economically viable level, respectively. Companies typically recapture the costs of technology investment through profits in the scale-up phase.

Acceleration of the lifecycle increases the importance of gaining market share in the commercialization phase, so that domestic manufacturers can seize the huge opportunities associated with the scale-up phase. In addition to achieving economies of scale, scale-up today requires achieving economies of scope by rapidly producing alternative versions of the same core product to satisfy diverse global customers.

Manufacturing capability gaps in the United States have led to the loss of substantial economic benefits. American researchers invented and commercialized industrial robots, for examples, with the first installation in a General Motors plant in 1961, but now the vast bulk of industrial robot production is done in Asia and Europe. The same pattern holds in energy storage and power generation, and in many other areas of technology. U.S.-based facilities no longer produce electronic displays for computer monitors, televisions, or handheld devices, such as the Kindle e-reader.[15]

The loss of such production capabilities affects U.S. national security as well as the national economy. The Defense Production Act Committee (DPAC) (see Appendix B) has identified a number of vital government needs that the United States is currently unable to meet via secure and reliable domestic production. DPAC analysis also revealed several key technologies for which an impending lack of U.S. production will likely create national security vulnerabilities. A sampling of specific vulnerabilities include aircraft landing gear; railcar components; large rotor disks for turbines; rocket engine parts; missile launch systems; unmanned aerial and ground vehicles; nuclear power components; aircraft fuselages; orbital vehicles; network routing and switching; optical data transport; advanced power electronics; low cost composites; and transmission conductors.

---

15. Gary P. Pisano and Willy C. Shih (2009), Restoring American competitiveness, *Harvard Business Review*, July.

3. Principles and Objectives of the National Strategy

Challenges to the competitiveness of U.S. advanced manufacturing have the potential to undermine the Nation's ability to create jobs, invent new industries, and protect itself from security threats in the 21st century. The United States should respond to these challenges with an *innovation policy for advanced manufacturing* that addresses (1) the complexity of public and private elements of modern manufacturing technologies, and (2) the ways in which these elements change over the technology lifecycle. By doing so, the Nation can, as Secretary of Commerce John Bryson put it recently, "build it here, sell it everywhere."[16]

Innovation Policy for Advanced Manufacturing

An innovation policy for advanced manufacturing must be capable of responding to a range of market failures. Markets sometimes fail to provide adequate incentives to manufacturers to make specific investments that would benefit the economy over the long-run. For example, high-risk technologies may be neglected if firms are uncertain that they will reap the benefits from investing in them. Workers' skills, too, may be subject to market failure. An employer may fear that a worker who receives a training benefit will leave employment before the training pays off for the employer, while the worker may lack the financial means to pay for the training on his or her own. Federal investments in research, technology, and education and training have helped to create and accelerate new industries, such as the semiconductor industry, in the past, when market forces alone would not have done so.

The traditional U.S. approach to innovation has particularly emphasized Federal investments in basic research. This kind of investment is an effective response to an important and continuing market failure. It continues to pay extraordinary dividends, creating opportunities for technological advances in support of U.S. economic growth and national security. The U.S. leads the world in science, and it must continue to strive to maintain this status.

However, the payoffs from Federal investments in basic research have not been fully captured by U.S.-based advanced manufacturing production facilities. In part because of the national strategies of our competitors and in part because of the increasing complexity of industrial technology, private investments in advanced manufacturing capabilities may not occur domestically unless the public sector makes strategic investments to address market failures in stages of the innovation process downstream from basic research. Many of these public investments must be coordinated with private co-investments to create assets such as worker skills, precompetitive technologies, and shared infrastructure, as the President's Council of Advisors on Science and Technology has emphasized.[17]

---

16. Secretary of Commerce John Bryson, "Remarks at U.S. Chamber of Commerce," December 15, 2011.
17. PCAST, *Advanced Manufacturing*.

Therefore, one core principle of an effective national strategy for advanced manufacturing is to take a cohesive approach to research, development, and deployment. The approach begins with more effective commercialization of federally-financed technologies developed as a result of basic research by national labs and universities. More fundamentally, it involves a stronger Federal emphasis on R&D that improves manufacturing processes and supports scale-up, in part by enabling better access to user facilities equipped with advanced manufacturing technologies. In addition, education and workforce training programs that help equip Americans to become highly-skilled manufacturing workers can be invaluable.

Strengthening the Industrial Commons

A cohesive approach to research, development, and deployment is particularly important to encourage investment in high-impact advanced manufacturing processes by small- and medium-sized enterprises (SMEs). SMEs comprise 86% of all manufacturing establishments and employ 41% of the U.S. manufacturing workforce, but they often lag in adopting new technologies.[18] Those SMEs that are highly innovative are often tightly linked to geographically-concentrated communities, or "industrial clusters." These clusters contain other innovative SMEs, larger firms that typically rely on smaller suppliers, academic and training institutions, and other supporting organizations.

Clusters sustain what innovation experts have called the "industrial commons."[19] Like the common pasture in medieval English villages in which livestock owned by many residents grazed together, the industrial commons provides many of today's manufacturers, particularly SMEs, with a chance to refresh their technological base from a set of shared knowledge assets and physical facilities. These common resources help to accelerate innovation and subsequent market penetration. Standards for system interfaces, measurement and test methods, and process control systems, for instance, allow firms within a supply chain or even firms that compete with one another to align their diverse product and process capabilities with opportunities to serve customers in different markets. Similarly, platform technologies in such areas as nanomaterial processing, additive manufacturing, advanced robotics, "smart" manufacturing, and green chemistry are assets that many firms in an industrial cluster can take advantage of, but that no single firm can typically produce on its own.

---

18. SMEs are defined as firms with fewer than 500 employees, consistent with the Small Business Administration's definition. Data are drawn from the Business Dynamics Statistics database, U.S. Census Bureau.
19. Pisano and Shih, *op. cit.*

3. PRINCIPLES AND OBJECTIVES OF THE NATIONAL STRATEGY

The industrial commons itself must be refreshed through continual investment to keep the knowledge base and physical infrastructure at the leading edge of technology. It may be difficult for the firms that would benefit from them to make these investments individually, since they cannot capture all of these benefits. Coordination of investments across firms is also subject to market failure. Such investments may also be perceived as too risky for the private sector to make because the gestation periods are long. The risk is heightened in many cases by the prospect that government-linked entities in other countries may make similar investments intended to benefit competing firms. The public sector, particularly Federal agencies, therefore has an important role to play as a co-investor with advanced manufacturers in the industrial commons.

Greater public co-investment in the industrial commons would yield a variety of benefits. It would make basic research investments more productive for the economy. It would strengthen the skills of workers who are in or may enter the advanced manufacturing workforce. And it would reduce private investment risk in product-specific applied R&D, leading to higher rates of technology adoption and more diverse product-specific innovations.[20]

Optimizing Federal Investments

A national strategy for advanced manufacturing should coordinate Federal investments across agencies more effectively. Currently, Federal investments in advanced manufacturing-related research, development, and deployment are largely provided through agency programs focused on accomplishing mission-specific goals. The resulting cross-agency portfolio may not adequately consider how certain investments would benefit multiple agencies and industries as well as contribute to economic competitiveness more generally. A *whole-of-government* innovation policy that complements the work of individual agencies and takes a portfolio view of advanced manufacturing investments would address this problem.

An effective national strategy for advanced manufacturing should also be responsive to private-sector needs. Public-private consultation about research, technology, and workforce needs at the national level will complement the consultative processes that already exist at the state and regional level throughout the country. The President's Advanced Manufacturing Partnership (AMP) represents an initial but important step toward improving private-sector engagement at the national level.

---

20. Tassey, Gregory (2010), "Rationales and Mechanisms for Revitalizing U.S. Manufacturing R&D Strategies", Journal of Technology Transfer 35 (June): 283-333.

Stakeholders and Objectives

The stakeholders in this national strategy include Federal agencies represented in the Interagency Working Group that authored this report and those represented on the Defense Production Act Committee (DPAC);[21] state, regional, and local public and private entities that support industrial clusters and associated partnerships; manufacturing enterprises of all sizes; a diverse set of institutions of higher education including research universities and community colleges; workers and unions; and the general public.

The Federal strategy for advanced manufacturing seeks to achieve five objectives, which are developed in more detail in the following five sections of the report. These objectives are interconnected; progress on any single one will make progress on the others easier. Coordination of Federal agencies through the NSTC will support an implementation process that pursues all objectives of the strategy in parallel.

Objective 1: *Accelerate investment in advanced manufacturing technology, especially by small and medium-sized manufacturing enterprises, by fostering more effective use of Federal capabilities and facilities, including early procurement by Federal agencies of cutting-edge products.*

Objective 2: *Expand the number of workers who have skills needed by a growing advanced manufacturing sector and make the education and training system more responsive to the demand for skills.*

Objective 3: *Create and support national and regional public-private, government-industry-academic partnerships to accelerate investment in and deployment of advanced manufacturing technologies.*

Objective 4: *Optimize Federal investment in advanced manufacturing by taking a portfolio perspective across agencies and calibrating accordingly.*

Objective 5: *Increase total U.S. public and private investments in advanced manufacturing R&D.*

---

21. The DPAC is described in more detail below and in Appendix C.

4. Accelerating Investment by Small and Medium-Sized Enterprises

Objective 1: *Accelerate investment in advanced manufacturing technology, especially by small and medium-sized manufacturing enterprises, by fostering more effective use of Federal capabilities and facilities, including early procurement by Federal agencies of cutting-edge products.*

This component of the strategy aims to improve the success of U.S.-based manufacturers, especially SMEs, in the commercialization and scale-up phases of the technology lifecycle. These phases are critical because they set firms on a path of sustainable job creation and profit generation. The focus on SMEs reflects their importance to the manufacturing sector and the difficulty they experience developing and adopting technological innovations.

We focus on three kinds of Federal actions that support advanced manufacturing investments: (a) increasing coordination of their investments related to advanced manufacturing with private and non-Federal investors, (b) purchasing products made by advanced manufacturers early in the scale-up phase, and (c) investing in targeted areas of critical importance to national security.

Private-Public Co-Investment

Economists have long known that investment opportunities that would benefit groups of firms are less likely to receive private-sector funding than opportunities that benefit individual firms. Investments that would strengthen the industrial commons by deepening an industry's knowledge base or creating industry-wide standards, for instance, are usually less attractive than those that promise to benefit a firm's bottom line directly. Yet, the former are necessary to achieve the latter. These challenges have been exacerbated by recent trends in advanced manufacturing, such as the rising complexity of global markets and the speed with which they change. Rising complexity expands the number of fields that have to be integrated to advance the state of the art, while the faster speed of change raises the risk that ideas and standards will become obsolete before they are fully implemented.

Joint public–private strategic planning and prioritization of opportunities to strengthen the industrial commons should help to overcome these challenges. Coordinating public and private investments related to advanced manufacturing can lead to benefits for both sectors, for instance, by ensuring that all key players participate in standard-setting and accelerating standards adoption.[22] Federal investments in applied research, and in demonstration facilities to which manufacturers have access, can accelerate commercialization of new advanced manufacturing processes and stimulate private investment in plant and equipment.

---

22. OSTP/USTR/OMB Memorandum on "Principles for Federal Engagement in Standards Activities to Address National Priorities," January 17, 2012, http://www.whitehouse.gov/sites/default/files/omb/memoranda/2012/m-12-08_1.pdf

The President's Advanced Manufacturing Partnership (AMP), announced in June 2011, is an important building block for this element of the strategy. AMP's mission is to identify opportunities for investments in R&D, precompetitive collaboration, and shared facilities and infrastructure that have the potential to transform advanced manufacturing in the United States.

Early Procurement

The Federal government is a major purchaser of products made by advanced manufacturers. Effective use of this purchasing power in support of agency missions can drive economies of scale and scope, especially for SME manufacturers who are then enabled to enter new markets and compete effectively on an international basis. The best-known historical example of effective early procurement is the semiconductor industry. Military buyers absorbed virtually the entire output of semiconductors in 1962. These large-scale purchases pushed prices down 96 percent in the next six years, driving a rapid increase of adoption by the commercial sector, which had grown to be more than 60 percent of the market over that period.[23]

Federal agencies are using early procurement in a variety of mission domains today. In energy, for example, manufacturers of such innovative products as biofuels, advanced batteries, and plug-in electric vehicles, many of them SMEs, are gaining scale economies and production experience as a result of early procurement. But there are substantial opportunities for key agencies such as the General Services Administration and the Department of Defense to expand their use of early procurement in order to accelerate innovation in advanced manufacturing.

Advanced Manufacturing for National Security

The authorities of the interagency Defense Production Act Committee (DPAC) may be drawn upon to strengthen shared infrastructure that supports advanced manufacturing by SMEs. The DPAC is charged with both evaluating current manufacturing capabilities essential to national defense and advising the President on how to resolve manufacturing capability gaps through use of Defense Production Act authorities, such as Title III. These authorities allow Federal investments (usually matched 1:1 with industry) to scale up economically viable means of production (see Appendix B).

Additionally, the Department of Defense (DOD) Manufacturing Technology (ManTech) Program enhances defense industrial base productivity. ManTech's advanced manufacturing investments are applied to drive down weapon system cost and delivery time as well as enhance system performance. This program contributes to improving domestic firms' production effectiveness at every tier of the supply chain. The ManTech program, which has a portfolio of projects in the Army, Navy, Air Force, Defense Logistics Agency (DLA) and Office of the Secretary of Defense (OSD), provides a crucial link between invention and deployment. These investments also enhance manufacturing efficiency and reduce industry's financial and technical risks. ManTech investments currently focus on electronics, metals, composites and advanced manufacturing enterprise (see Appendix C).

---

23. Morris, Peter Robin, *A History of The World Semiconductor Industry* (London: Peter Peregrinus, 1990), p. 75.

4. ACCELERATING INVESTMENT BY SMALL AND MEDIUM-SIZED ENTERPRISES

Metrics and Key Implementing Agencies for Objective 1

Metrics for short-term progress on Objective 1 could include: (a) documented instances of commercialization and scale-up by firms participating in Federal programs, both manufacturers as a whole and SME manufacturerss (b) use of Federal user facilities by advanced manufacturing firms, and (c) industry matching of Federal DPA Title III investments.

Suggested metrics for long-term progress on Objective 1 include: (a) growth of advanced manufacturing production capacity overall and by SMEs, (b) share of SMEs reporting innovation activity in the manufacturing sector as measured by the Business R&D and Innovation Survey (BRDIS) of the National Science Foundation (NSF),[24] and (c) growth in the U.S.-produced share of the global advanced technology products market.

The primary Federal agencies that should implement actions under Objective 1 include the Departments of Commerce, Defense, Energy, Homeland Security, and Transportation, and the General Services Administration (GSA), National Aeronautics and Space Administration (NASA) and the Small Business Administration (SBA).

---

24. National Science Foundation (2011) *NSF Releases New Statistics on Business Innovation,* NSF11-300, http://www.nsf.gov/statistics/infbrief/nsf11300/nsf11300.pdf

5. Strengthening Workforce Skills

Objective 2: *Expand the number of workers who have skills needed by a growing advanced manufacturing sector and make the education and training system more responsive to the demand for skills.*

Unskilled labor was once the mainstay of the manufacturing labor force. As advanced manufacturing supersedes traditional manufacturing, and domestic manufacturers deepen their investment in advanced technologies, the skill requirements for manufacturing jobs are rising. Manufacturing employers perceive a skills gap: 67% of companies surveyed recently by an industry association reported moderate to serious shortages in the availability of qualified workers, even in a period of elevated general unemployment. Certain sectors, such as aerospace/defense and life sciences/medical devices, reported much higher levels of skilled-worker shortages.[25]

Education and training that anticipates and satisfies the skill requirements of advanced manufacturers, while remaining broadly consistent with long-term projections of labor demand, is a key component of this national strategy. Increasing the private sector's confidence in the availability of a skilled advanced manufacturing workforce creates incentives for domestic investment (see Objective 1). These programs should be targeted particularly toward the workforce needs of SMEs. As more advanced manufacturing technology is deployed, on-the-job training becomes more expensive and difficult for companies, especially SMEs, to provide.

Federal actions under this objective should include such efforts as (a) support for the coordination of state and local education and training curricula with advanced manufacturing skill-set requirements, and (b) expanded support for advanced manufacturing career and technical education programs spanning secondary and postsecondary levels, and apprenticeship opportunities through regional partnerships and industrial cluster programs.

---

25. Deloitte Consulting LLP, Manufacturing Institute (2011), *Boiling Point? The skills gap in U.S. manufacturing.*

The Changing Manufacturing Workforce

The shift from traditional to advanced manufacturing is occurring in the context of a substantial shift in the demographics of the manufacturing workforce. Approximately 2.8 million manufacturing workers (nearly 25%) are now 55 years of age or older.[26] The need to replace these workers as they retire may add to emerging demand for advanced manufacturing workers. In the long term, education and training programs must span from "cradle-to-career" and be responsive to the skill demands of advanced manufacturing employers. Federal programs in cooperation with state and local partners should target (a) separating military personnel and recent veterans, unemployed workers, and employed workers needing to augment their skills in the near-term, (b) prospective workers who will soon enter the workforce, and (c) K–12 students to proactively develop the next-generation of workers.

Better Training for Today's Advanced Manufacturing Workers

The Federal Government is already seeking to adjust current programs that assist state and local public and private efforts to develop and maintain a competitive workforce for advanced manufacturing. Relevant agencies may prioritize advanced manufacturing within workforce development grant programs. They may also identify and disseminate workforce development "best practices" for advanced manufacturing. Many of these practices arise from competitive grants funded by the Department of Labor's Employment and Training Administration (DOL/ETA). Such efforts should be expanded. Among the programs that could increase priority for advanced manufacturing are H-1B Technical Skills Training Grants,[27] the Jobs and Innovation Accelerator Challenge, and Trade Adjustment Assistance Community College and Career Training (TAA/CCCT).

This approach could be complemented by a new emphasis on advanced manufacturing in the promotion of secondary-postsecondary career pathways[28] by the Department of Education's Office of Vocational and Adult Education (ED/OVAE). The President's Fiscal Year (FY) 2013 Budget proposes $8 billion for the Departments of Education and Labor to support state and community college partnerships with businesses to build the skills of American workers in growing industries, such as advanced manufacturing.

---

26. Bureau of Labor Statistics, *2010 Current Population Survey*.
27. H-1B Technical Skills Training Grants provide education, training and job placement assistance related to high-growth fields in which employers are currently using the H-1B nonimmigrant visa program to hire foreign workers, including advanced manufacturing, These grants are funded through fees paid by employers under the H-1B program.
28. These are referred to as "programs of study" in the Perkins CTE legislation.

Education and Training for Tomorrow's Workers

State and local vocational and apprenticeship training programs supported by the Federal Government strengthen workers' skills. For example, the National Institute of Standards and Technology (NIST) Manufacturing Extension Partnership (MEP) and DOL/ETA support public–private partnerships that establish registered apprenticeship programs in advanced manufacturing. With appropriate input from industry and professional associations, DOL/ETA will seek to ensure that registered apprenticeship programs target needs and gaps in today's advanced manufacturing workforce.

The Advanced Technological Education (ATE) program of the National Science Foundation supports community colleges working in partnership with industry, economic development agencies, workforce investment boards, and secondary and other higher education institutions to respond to industry needs for highly qualified manufacturing technicians. Since the inception of the program in 1994, 265 manufacturing awards have been made totaling $205 million. ATE projects and centers are educating technicians in a range of fields, including nanotechnologies and microtechnologies, rapid prototyping, biomanufacturing, logistics, and alternative fuel automobiles. More details on these and other advanced manufacturing workforce programs can be found in Appendix D.

Another set of partnerships that is being leveraged to improve the training of tomorrow's workers are those organized by the National Association of Manufacturers (NAM) and the Society of Manufacturing Engineers. NAM has helped foster the Manufacturing Skills Certification System, a broad-based partnership of national organizations seeking to establish a set of credentials that apply to all subsectors of manufacturing.[29] This system draws on the advanced manufacturing competency model developed by the Department of Labor, which identifies the knowledge and skills required to be effective in advanced manufacturing occupations (see Appendix E). Credentials issued within this framework will be nationally portable and industry-recognized. They will help to ensure a smart, safe, and sustainable advanced manufacturing workforce. The Society of Manufacturing Engineers provides extensive training opportunities and associated certifications under this partnership. In June 2011, President Obama drew on the work of NAM, the Society of Manufacturing Engineers, and others when he announced an effort to help 500,000 community college students obtain credentials for advanced manufacturing as part of the Administration's Skills for America's Future initiative.

---

29. The Manufacturing Institute (2010) NAM-Endorsed Manufacturing Skills Certification System, see http://www.themanufacturinginstitute.org/Education-Workforce/Skills-Certification-System/Skills-Certification-System.aspx.

5. STRENGTHENING WORKFORCE SKILLS

Educating the Next Generation

A strong science, technology, engineering, and mathematics (STEM) emphasis is needed to prepare students for a variety of post-secondary educational options and a wide range of career opportunities, including careers in advanced manufacturing.[30] Unfortunately, many students who are inclined toward careers as advanced manufacturing technicians who seek to complete a career and technical certificate or community college degree have not received STEM coursework sufficient to succeed in today's advanced manufacturing environment. The President's Educate to Innovate campaign aims to improve participation and performance in STEM education in partnership with leading companies, foundations, and scientific and professional societies.

The perception among some that careers in manufacturing are unattractive and unstable discourages some talented students from seriously exploring them. Federal agencies should consider fostering efforts that aim to transform this perception and consider supporting development of communications materials that accurately depict the opportunities and excitement of 21st century manufacturing.

In recent years, there has been a rapidly growing grassroots movement of "Makers" who are engaged in "do it yourself" projects involving electronics, 3-D printing and robotics. These hands-on projects inspire young people to excel in STEM and can also get them interested in advanced manufacturing. Some agencies, such as DARPA through its Mentor program (see Appendix D), are supporting the Maker movement. More agencies should support Making, in collaboration with the private sector, non-profits, foundations and skilled volunteers.

Another major issue related to preparing students for advanced manufacturing careers is the need to supplement traditional academic education with the development of applied expertise. The Federal Government should help state and local efforts to develop this applied expertise by supporting new manufacturing pre-apprenticeship programs, strengthening existing educational partnerships between community colleges and local industry, and other measures.

Feedback from these programs will help align the curricula of feeder high schools and adult education programs within the service area of each community college. Alignment of these curricula with four-year degree curricula will help provide more attractive career pathways in advanced manufacturing. A critical example is the pathway from a two-year degree to an eventual four-year degree, which enables highly-skilled technical workers to pursue additional education and higher-paying careers.

---

30. The National Science and Technology Council's Committee on Science, is carrying out a comprehensive effort to strengthen the Nation's STEM education system. The most recent public report is NSTC, *The Federal Science, Technology, Engineering, and Mathematics (STEM) Education Portfolio (2011).*

Metrics and Key Implementing Agencies for Objective 2

The following metrics may be applied to assess progress on Objective 2 in the short-term: (a) industry partnerships with state and local career and technical education providers to specify and implement advanced manufacturing workforce skill sets, and (b) number of individuals earning industry recognized-credentials through career and technical education, vocational training, and apprenticeships programs.

Suggested metrics for assessing long-term progress on Objective 2 include: (a) employment placement rates for graduates from state and local career and technical education, vocational training and apprenticeship programs, and (b) employment levels in highly-skilled manufacturing occupations.

The primary Federal agencies that should implement actions under Objective 2 include DOL/ETA, ED OVAE, and NSF.

6. Creating Partnerships

Objective 3: *Create and support national and regional public-private, government-industry-academic partnerships to accelerate investment in and deployment of advanced manufacturing technologies.*

The acceleration of innovation for advanced manufacturing requires bridging a number of gaps in the present U.S. innovation system. Academic researchers working on problems of importance to advanced manufacturing must communicate more effectively with their counterparts in industry. Federal investments in advanced manufacturing technologies and capabilities must align more fully with similar investments by states and regions and by the private sector. Partnerships among diverse actors, varying by location and objective, are a keystone of our strategy to bridge these gaps.

Actions under this objective will focus more of the Federal advanced manufacturing investment portfolio on partnership activities, especially those that yield benefits for SMEs. These actions include (a) facilitating SME engagement through partnerships, and (b) expanding investments in public–private partnerships in the advanced manufacturing industrial commons.

SME Engagement through Partnerships

SMEs rely on the industrial commons to achieve economies of scale and economies of scope during commercialization and scale up. The production infrastructure that resides in the commons helps SMEs meet diverse market demands through "mass customization" capabilities that use advanced manufacturing technologies to rapidly adjust product attributes while maintaining high quality and low unit cost. Industrial clusters and industry segments that can draw upon a vibrant commons have lower entry barriers for SMEs because these SMEs have easier access to widely-available expertise in advanced manufacturing technologies, innovation, and commercialization.

Federal investments should be better targeted to strengthen the industrial commons, in large part by supporting cross-sectoral partnerships. Partnerships that involve academic institutions, manufacturers, industry associations, and supporting organizations may be structured to produce ideas and capabilities that support commercialization and scale-up activities across a wide swath of firms. A focused strategy for the commercialization of technologies that draw on Federally-funded research, which was recently promulgated by the Administration, will help as well.[31]

The National Design Engineering and Manufacturing Consortium (NDEMC) is a good example of a project that strengthens the industrial commons. Funded by the Economic Development Administration in the Department of Commerce (along with other Federal agency, private, state, and university partners that provide financial or technical assistance), NDEMC enables SMEs to develop and test their products using advanced modeling and simulation tools that have historically only been available to large companies. NDEMC is designed to exploit regional commonalities among supply chain industries and its operational framework is expected to be self-sustaining, scalable, and transferable to other industry sectors and regions across the U.S.

---

31. Presidential Memorandum (October 2011) *Accelerating Technology Transfer and Commercialization of Federal Research in Support of High-Growth Businesses.*

Manufacturing Demonstration Facilities co-funded by DOE with private partners provide physical and virtual tools for rapidly prototyping advanced materials and manufacturing processes within targeted technical areas. These facilities host a collaborative, shared infrastructure that facilitates rapid dissemination of these new technologies. The technical areas already targeted by DOE through this program include carbon fiber and additive manufacturing, with additional areas envisioned in DOE's current plans. MDFs support regional manufacturing ecosystems by providing small- and medium-sized enterprises access to innovative tools and resources that would otherwise be cost-prohibitive and help reduce energy intensity, create lower cost production pathways, and enhance the competitiveness of U.S. advanced manufacturing industries.

Cluster-Based Partnerships

Regional industrial clusters provide a fruitful setting for investments in partnerships. These clusters dramatically improve technology platform development, diffusion of knowledge, and subsequent scale-up. Additional synergies achieved through regional clusters include coordinated strategic planning, complementary asset sourcing and risk pooling within clusters, and co-located supply chains. Clusters provide vital resources for SMEs. Shared production facilities, for example, provide SMEs with both production experience and the knowledge to develop scale-up plans.[32]

The nano-technology cluster that is emerging in Albany, New York, provides an example of the benefits of such partnerships. The State of New York assembled key stakeholders, co-invested with them, and anchored the effort at the State University of New York in Albany. Large manufacturers, such as IBM, along with the semiconductor industry consortium Sematech, made major commitments, and the cluster has given birth to a number of promising start-ups.[33]

The Federal Government is making cluster-based investments that bring together educational and research organizations, state and regional economic development authorities, and the private sector to conduct proof-of-concept and commercialization activities. For instance, the Jobs and Innovation Accelerator Challenge, which is led by the Economic Development Administration (EDA) in the Department of Commerce in partnership with other agencies, such as the Department of Energy (DOE), NIST, and the Small Business Administration (SBA), will run a competition focused on advanced manufacturing in fiscal 2012.

Federal agencies will also seek to support novel approaches in this context to enhance competitiveness in advanced manufacturing. The E3 Initiative, a framework that connects EPA, NIST, DOE, DOL, and SBA to local and regional groups of SMEs to deploy advanced manufacturing processes to reduce waste and improve operational efficiency, is an example of such a novel approach.[34]

---

32. Gregory Tassey (2011) *Beyond the Business Cycle: The Need for a Technology-Based Growth Strategy* (http://www.nist.gov/director/planning/upload/beyond-business-cycle.pdf). For a broader statement of cluster theory and empirical evidence, see Mercedes Delgado, Michael E. Porter, and Scott Stern, "Clusters, Convergence, and Economic Peformance," Institute for Strategy and Competitiveness, Harvard Business School, March 2011(http://www.isc.hbs.edu/pdf/DPS_Clusters_Performance_2011-0311.pdf).
33. Board on Science, Technology, and Economic Policy, National Research Council, Growing Innovation Clusters for American Prosperity: *Summary of a Symposium* (National Academies Press, 2011).
34. E3: Economy, Energy, Environment: Supporting Manufacturing Leadership through Sustainability, see www.e3.gov.

Metrics and Key Implementing Agencies for Objective 3

The following metrics may be used to measure short-term progress on Objective 3: (a) number of Federally-funded public-private partnerships that aim to accelerate innovation in advanced manufacturing, (b) total scale of investment in such partnerships and (c) ratio of private to public funding within these partnerships.

Suggested metrics for long-term progress on Objective 3 include: (a) publications and patents related to advanced manufacturing generated by basic research activities associated with partnerships,[35] (b) number of products commercialized by U.S.-based firms engaged in advanced manufacturing partnerships and total global sales of such products, and (c) advanced manufacturing location quotients for regions in which advanced manufacturing partnerships are active.[36]

The primary Federal agencies that should implement actions under Objective 3 include DOD, DOE, EDA, NIST, NSF, and SBA.

---

35. Due to the nature of basic research, metrics for long term progress on this objective will draw on best assessment practices from existing industry-university research center programs (e.g. NSF Engineering Research Centers, NSF Industry / University Collaborative Research Centers) as well as the use of emerging Science of Science Policy tools (e.g. Star Metrics) as appropriate.

36. Location Quotients (LQs) are ratios that allow an area's distribution of employment by industry to be compared to a reference or base area's distribution. A high location quotient means that an industry is particularly concentrated in an area. See BLS help page: http://www.bls.gov/help/def/lq.htm.

7. Coordinating Federal Investments

Objective 4: *Optimize the Federal advanced manufacturing investment by taking a portfolio perspective across agencies and adjusting accordingly.*

A number of Federal agencies make research, development, and deployment investments that directly or indirectly benefit advanced manufacturers in the United States. These investments are typically made by agencies independently pursuing their statutory missions. The benefits from these investments can be augmented by analyzing them as a portfolio and adjusting agency investment strategies to reflect this analysis. Such adjustments can be made without compromising agencies' responsiveness to their individual missions. In fact, more efficient and faster development of new technology platforms will enhance the achievement of agency missions.

Actions under this objective include: (a) coordinating Federal agency investments in the industrial commons, and (b) targeting and balancing investments in advanced materials, broad production technology platforms, advanced manufacturing processes, and design and data infrastructure.

The Federal Advanced Manufacturing Investment Portfolio

The Federal Government makes investments in advanced manufacturing R&D and in some kinds of plants and equipment that private industry typically avoids. These high-risk investments help to position promising, nascent technologies (1) for broad adoption and commercialization, or (2) to meet essential DPAC-identified national security needs (see Objective 1). These investments can be grouped into the following four categories:

- Advanced Materials
- Production Technology Platforms
- Advanced Manufacturing Processes
- Data and Design Infrastructure

These four categories encompass the entire spectrum of relevant advanced manufacturing technology investments, from materials that will go into the products of the future to the factory and enterprise processes that enable production. Examples of current Federal investments in these areas are included in Appendix F. Descriptions of the four categories follow. Creating a coordinated Federal portfolio of investments across these four categories will increase the global competitiveness of U.S. manufacturing and help to create a fertile domestic environment for innovation.

Advanced Materials

Materials are the building blocks of every physical product. Traditional materials such as steels, metals, plastics, and ceramics have been improved routinely by materials scientists and chemists. These improvements were vital to many of the significant technological developments of the last century. Newer nanoscale, biological, smart, and composite materials will enable technological breakthroughs in the coming century. Some of these breakthroughs will transform existing industries. Others will spawn entirely new industries.

The Materials Genome Initiative is a key component of the Federal advanced manufacturing investment portfolio. The Departments of Energy and Defense, along with NSF, NIST, and private, academic, and state partners, seek through this initiative to cut the time it takes to develop advanced materials, such as those needed to make vehicles much lighter or dramatically increase the energy density of batteries. The initiative is building the infrastructure and providing the training needed for Americans to discover, develop, manufacture, and deploy advanced materials in a more expeditious and economical way.

Product Technology Platforms

Product technology platforms are the basic means of making products. Manufacturing engineers developed critical improvements in production technologies for components such as display screens, computer chips, batteries, nanoelectronic devices, gears, cases, and bolts. These improvements enabled critical advancements in diverse families of products, such as cell phones, computers, televisions, air bags, automobiles, and satellites. Future product technology platform innovations will enable more flexible and smarter production operations.

Sustainable nanomanufacturing, a signature initiative of the multi-agency National Nanotechnology Initiative, exemplifies Federal efforts to support development of product technology platforms that will provide the basis for future manufacturing industries. These industries will need methods to efficiently assemble products that integrate billions of nanoscale devices with disparate functions. Current manufacturing methods such as those used in the semiconductor industry will not be economical at these scales. The initiative is therefore pursuing radically new technical approaches that promise to lead to flexible, "bottom-up" or "topdown/bottom-up" continuous assembly methods.

Advanced Manufacturing Processes

Manufacturing processes turn materials into components and components into products. Manufacturing process innovations such as semiconductor wafer fabrication, composite structure processing, and bio-production of pharmaceuticals have enabled the introduction of new components and products, such as smartphones, military aircraft, and flu vaccines, in both new and existing markets. Emergent examples of advanced manufacturing processes include additive manufacturing, composite structure manufacturing, and bio-manufacturing.

The Innovative Manufacturing Initiative of the Department of Energy is one element of the Federal portfolio that is seeking to accelerate advanced manufacturing process innovation. It is funding cost-shared R&D of processes that have the potential to significantly reduce energy and carbon intensity over the coming decades. By doing so, the initiative aims to revitalize existing manufacturing industries as well as support the development of emerging industries. In a similar fashion, the Biorefinery Assistance Program of the Department of Agriculture is furthering the development and commercialization of advanced biofuels by providing loan guarantees to first-of-a-kind advanced biofuel biorefineries.

Data and Design Infrastructure

The ability to integrate all three types of innovation described above -- advanced materials, product technology platforms, and advanced manufacturing processes -- requires effective use of large quantities of manufacturing data and sophisticated design knowledge. Decisions that draw on such data and knowledge to achieve this integration can reduce the time for process improvements to be instituted and for products to reach customers. Data and design infrastructure, such as modeling standards for products and processes, interface standards for functional and physical connections between components, and mechanisms for flexible system integration, make integrated decision-making of this sort far easier and more accessible, especially for SMEs.

The manufacturing initiative of the Defense Advanced Research Project Agency (DARPA) within the DOD includes a number of elements that will strengthen the advanced manufacturing data and design infrastructure. For instance, the initiative seeks to develop design tools that enable correct-by-construction design through model-based verification for complex cyber-physical defense systems. It also seeks to create a bit-stream-configurable "foundry" capable of rapid switch-over across product variants and families leading to mass customization.

Cross-Cutting Agency Investments

Agencies can leverage their funding to invest in projects that they may not otherwise have the resources to support on their own by collaborating with other agencies on joint funding solicitations and co-funding of projects. These cross-cutting manufacturing investments should be used to strengthen the industrial commons in ways that would benefit all the participating agencies (and their stakeholders). For example, agencies could co-fund advanced materials and new design methods that dramatically decrease time to market for innovations. These targeted elements are typically beyond the purview of any one agency or private-sector entity, but are collectively viewed as critical to advancing key national interests. Individual agencies may supplement these investments with additional investments that focus specifically on their agency-unique missions or needs. The recently created Advanced Manufacturing National Program Office provides an administrative mechanism to coordinate and manage such investments.

Metrics and Key Implementing Agencies for Objective 4

Progress toward Objective 4 in the short-term might be measured by: (a) development and implementation of a framework for managing the whole-of-government portfolio, and (b) number and scale of multi-agency advanced manufacturing funding solicitations.

Suggested metrics for long-term progress on Objective 4 include: (a) balance of Federal advanced manufacturing R&D investment across portfolio dimensions, including character of work (i.e. basic research, applied research, demonstration facilities, etc.) and (b) accelerated time-to-market of new advanced manufacturing processes and products.

The primary Federal agencies implementing actions under Objective 4 should include DOD, DOE, NIST, and NSF.

8. Raising National Investment in Advanced Manufacturing R&D

Objective 5: *Increase total U.S. public and private investments in advanced manufacturing R&D.*

Objectives 1 through 4 are essential components of a cohesive national strategy that will increase the payoff from each dollar of Federal investment. They call for leveraging and coordinating Federal investment through partnerships with other levels of government and other public institutions, such as universities, as well as with manufacturers and industry and professional associations. They call for the creation of complementary assets, such as worker skills and industrial know-how, that will fill major gaps in the U.S. national innovation system.

However, actions to meet Objective 1 through 4 will not take full advantage of the present opportunity to sustain and strengthen the Nation's advanced manufacturing sector. To do that, the Nation must raise its level of investment in R&D as well, as the President has called for. The complementary nature of private and public investments suggests that an increase in one sector will be followed by an increase in the other.

Actions under this objective include (a) enhancing and making permanent the Federal Research and Experimentation (R&E) tax credit in order to expand the scope of activities covered and benefit a larger number of manufacturers, and (b) increasing Federal investment for advanced manufacturing R&D.

R&E Tax Credit

Federal tax policy has long provided incentives for private investments in Research and Experimentation (R&E).[37] The policy is well-rooted in economic theory. Private R&D investments create new knowledge that may not pay off for the firm making them, but may turn out to be useful to other firms or to society as a whole, a possibility that may discourage them from making such investments in the first place. The R&E tax incentive helps to overcome this barrier. Further, even applied R&D, which is the focus of industry investment, is risky along both technical and market dimensions compared with other categories of investment. Industry, therefore, systematically under invests in R&D.

However, the structure and size of the current U.S. R&E tax credit is ineffective. Although the U.S. was one of the first nations to enact such a credit, other nations have surpassed the U.S. over the years by offering more attractive provisions. Sixteen industrialized countries that are members of the OECD, for instance, offer tax credits that provide a greater incentive than that of the United States.[38] Such incentives are increasingly important in a world in which there is growing international competition for corporate R&D investment. The President's 2013 Budget proposes enhancing and making permanent the R&E tax credit.

---

37. The statutory definition of "research and experimentation" differs from the colloquial "research and development" in that it excludes spending that supports development of a specific product. See Science and Engineering Indicators 2012, p. 4-36.
38. OECD, *Science, Technology and Industry: Outlook 2008*, http://www.sourceoecd.org/9789264049918.

8. RAISING NATIONAL INVESTMENT IN ADVANCED MANUFACTURING R&D

Federal Investment

The Federal Government plays an important role in investing in R&D that helps to foster the development of advanced manufacturing processes and products. The President's Fiscal Year (FY) 2013 Budget provides $2.2 billion for Federal advanced manufacturing R&D—an increase of more than 50% over the 2011 level—at NSF, DOE, NIST, and other agencies. Important research funded by the Budget includes:

- An additional $86 million above the FY 2012 enacted level for the National Institute of Standards and Technology (NIST) laboratories to expand research in areas such as smart manufacturing, biomanufacturing and nanomanufacturing, as well as $21 million for the Advanced Manufacturing Technology Consortia program, a public-private partnership that will develop road maps for long-term industrial research needs and fund research at universities, government laboratories, and businesses directed at meeting those needs.

- $290 million – more than double the amount in FY 2012 – for the Advanced Manufacturing Office within DOE's Office of Energy Efficiency and Renewable Energy. This program is expanding its R&D activities on innovative manufacturing processes and advanced industrial materials that will enable U.S. companies to cut the costs of manufacturing by using less energy, while improving product quality and accelerating product development. The Budget also continues to support the development of competitive new manufacturing processes for advanced vehicles, biofuels, solar energy and other new clean energy technologies, to help ensure that the technologies invented here are manufactured here.

- An increase of $39 million above the FY 2012 enacted level, for basic research at the National Science Foundation targeted at developing revolutionary new manufacturing technologies in partnership with other Federal agencies and the private sector.

These investments will enhance the competitiveness of the Nation's manufacturing sector by spurring the development of new manufacturing technologies, processes, and materials.

Metrics and Key Implementing Agencies for Objective 5

Potential indicators for measuring progress on Objective 5 over the short term include: (a) enactment of the Administration's tax reforms, including enhancing and making permanent the R&E tax credit, and (b) funding levels for Federal advanced manufacturing R&D.

Metrics for assessing progress on Objective 5 over the long term include: (a) scale of use of the R&E tax credit, and (b) funding levels for Federal advanced manufacturing R&D.

The primary Federal agencies implementing actions should include the Department of the Treasury, DOD, DOE, NIST, and NSF.

Appendix A: America COMPETES Reauthorization Act of 2010 Section 102.

COORDINATION OF ADVANCED MANUFACTURING RESEARCH AND DEVELOPMENT.

(a) INTERAGENCY COMMITTEE.-The Director shall establish or designate a Committee on Technology under the National Science and Technology Council. The Committee shall be responsible for planning and coordinating Federal programs and activities in advanced manufacturing research and development.

(b) RESPONSIBILITIES OF COMMITTEE.- The Committee shall -

1. coordinate the advanced manufacturing research and development programs and activities of the Federal agencies;

2. establish goals and priorities for advanced manufacturing research and development that will strengthen United States manufacturing;

3. work with industry organizations, Federal agencies, and Federally Funded Research and Development Centers not represented on the Committee, to identify and reduce regulatory, logistical, and fiscal barriers within the Federal government and State governments that inhibit United States manufacturing;

4. facilitate the transfer of intellectual property and technology based on Federally supported university research into commercialization and manufacturing;

5. identify technological, market, or business challenges that may best be addressed by public-private partnerships, and are likely to attract both participation and primary funding from industry;

6. (encourage the formation of public-private partnerships to respond to those challenges for transition to United States manufacturing; and

7. develop, and update every 5 years, a strategic plan to guide Federal programs and activities in support of advanced manufacturing research and development, which shall-

    A. specify and prioritize near-term and long-term research and development objectives, the anticipated time frame for achieving the objectives, and the metrics for use in assessing progress toward the objectives;

    B. specify the role of each Federal agency in carrying out or sponsoring research and development to meet the objectives of the strategic plan;

    C. describe how the Federal agencies and Federally Funded Research and Development Center supporting advanced manufacturing research and development will foster the transfer of research and development results into new manufacturing technologies and United States based manufacturing of new products and processes for the benefit of society to ensure national, energy, and economic security;

D. describe how Federal agencies and Federally Funded Research and Development Centers supporting advanced manufacturing research and development will strengthen all levels of manufacturing education and training programs to ensure an adequate, well-trained workforce;

E. describe how the Federal agencies and Federally Funded Research and Development Centers supporting advanced manufacturing research and development will assist small- and medium sized manufacturers in developing and implementing new products and processes; and

F. take into consideration the recommendations of a wide range of stakeholders, including representatives from diverse manufacturing companies, academia, and other relevant organizations and institutions,

(c) REPORT.-Not later than 1 year after the date of enactment of this Act, the Director shall transmit the strategic plan developed under subsection (b)(7) to the Senate Committee on Commerce, Science, And Transportation, and the House of Representatives Committee on Science and Technology, and shall transmit subsequent updates to those committees as appropriate.

Appendix B: The Defense Production Act Committee (DPAC)

The Defense Production Act Committee (DPAC) supports a whole-of-government approach to manufacturing production policy. The DPAC is an interagency body comprised of the majority of the heads of the Federal Government's Departments and Agencies. The Committee's staff conducts assessments of the U.S. industrial base to identify risks within supply chains deemed essential to multiple Department or Agency missions, and provide recommendations to the President for appropriate mitigation. The DPAC also continually reviews DPA authorities and their applications to provide recommendations, as necessary, to Congress and the President on appropriate legal or regulatory modifications.

The DPAC was established by Congress in 2009 to advise the President on the effective use of a law enacted in 1950 (50 U.S.C. App. § 2171 et seq.). The President has directed the Secretaries of Defense and Homeland Security to rotate annually as DPAC Chairperson. DPAC assessments are expected to inform joint-Departmental use of a revolving "DPA Fund," established under Title III of the law to expand the "productive capacity and supply" of important domestic industries. Title III investments are designed to support industrial/technological capabilities that are commercially viable and essential to military production, energy production or construction, military or critical infrastructure assistance to any foreign nation, homeland security, stockpiling, space, and any directly related activity as well as emergency preparedness activities conducted pursuant to title VI of The Robert T. Stafford Disaster Relief and Emergency Assistance Act [42 U.S.C. § 5195 et seq.] and critical infrastructure protection and restoration. Title III assistance is only made available if U.S. industry could not otherwise be expected to provide the needed capability in a timely manner. Moreover, under the law's Statement of Policy in subsection 2(b), "in providing United States Government financial assistance to correct a domestic industrial base shortfall, the President should give consideration to the creation or maintenance of production sources that will remain economically viable after such assistance has ended."

Under Executive Order, the DPA Fund is managed by the DOD, and promotes interagency collaboration with industry on capital expenditures such as retrofits, machine tool acquisitions, or wholesale plant construction. Monies are transferred to the Fund and expended according to requirements established by an Integrated Product Team (IPT), comprised of DOD officials and representatives of the Departments/agencies contributing to the Fund. According to the DPA Section 2B (1), all Departments and agencies responsible for defense procurement are required to ensure the adequacy of productive capacity and supply. The DPAC supports this key interagency responsibility by providing industrial base assessments and targeted studies of industry areas on behalf of the member Departments and Agencies.

Appendix C: The Department of Defense (DoD) Manufacturing Technology (ManTech) Program

The DoD ManTech Program develops advanced manufacturing technologies and processes for the affordable, timely production and sustainment of defense systems. In close partnership with industry, the program impacts all phases of system development, acquisition and sustainment by developing, maturing, and transitioning key advanced manufacturing technologies. Investments are focused on those technologies that have the greatest benefit to the Warfighter and are balanced to support transition of emerging technologies, improvements to existing production enterprises, and strengthening the U.S. Industrial base. ManTech has a long history of delivering critical and "game changing" advanced manufacturing technologies and processes, such as numerically controlled machines, carbon fiber composites, microelectronics fabrication, advanced radars, laser guided munitions, lean production methods, advanced optics, and advanced soldier body armor. Many defense manufacturing technologies have 'spun-off" into commercial markets and resulted in large economic advances for the U.S.

ManTech was established in 1956 and is codified in Title 10, United States Code, to increase national security "…through the development and application of advanced manufacturing technologies and processes that will reduce the acquisition and supportability costs of defense weapon systems and reduce manufacturing and repair cycle times across the lifecycles of such systems." ManTech investments support advanced manufacturing through research and development of processing and fabrication solutions to support emerging technologies as well as enterprise-level initiatives. Current and planned initiatives are described in the 2009 DoD ManTech Program Strategic Plan and include strong support for a highly connected and collaborative manufacturing enterprise, a deep, institutional focus on manufacturing process maturity, and healthy and resilient manufacturing infrastructure and workforce.

The Army, Navy, Air Force, and Defense Logistics Agency (DLA) each manage robust ManTech programs. In addition, the Office of the Secretary of Defense (OSD) has responsibility for managing the Defense-wide Manufacturing Science & Technology (DMS&T) portfolio of R&D investments. The directors and senior managers of these programs coordinate through the congressionally chartered Joint Defense Manufacturing Technology Panel (JDMTP), comprised of one principal representative from each program. The JDMTP identifies and integrates ManTech R&D requirements, conducts joint program planning, and develops joint strategies for advanced manufacturing. ManTech investments are jointly reviewed via a taxonomy of technology focus areas assigned to subpanels. The current subpanel technology focus areas are Electronics, Metals, Composites and Advanced Manufacturing Enterprise.

The position of Deputy Assistant Secretary of Manufacturing and Industrial Base Policy (DASD(MIBP)) was established by the 2011 National Defense Authorization Act (NDAA) and has oversight responsibility for the DoD ManTech Program on behalf of the Secretary of Defense. The DASD(MIBP) is responsible for ensuring that the functions of "manufacturing" and "industrial policy" are effectively integrated and coordinated across the department, and the DoD ManTech Program is an important tool supporting such integration and coordination.

In summary, the DoD ManTech program provides the crucial link between technology invention and robust applications across the U.S. defense industrial base by maturing and validating advanced manufacturing technologies to support low-risk implementation by industry and DoD facilities, such as depots.

Appendix D: Examples of Federal Support for the Advanced Manufacturing Workforce

Defense Advanced Research Projects Agency (DARPA)

DARPA's Manufacturing Experimentation and Outreach (MENTOR) effort engages high school students in collaborative distributed manufacturing and design. MENTOR develops next-generation system designers and advanced manufacturing innovators. MENTOR will (1) develop of user-friendly, open-source tools enabling collaborative distributed design and manufacturing across hundreds of sites and thousands of users through conventional social network media and (2) deploy digitally programmable manufacturing equipment to 1,000 high schools. Clusters of schools will compete in prize-based design and manufacturing challenges to develop of cyber-electromechanical systems of moderate complexity, such as go-carts, mobile robots, or small unmanned aircraft.

Department of Commerce, National Institute of Standards and Technology (NIST)

In June 2011, President Obama announced new commitments by the private sector, colleges, and the Manufacturing Institute of the National Association of Manufacturers (NAM) to train and certify 500,000 community college students through the NAM-Endorsed Manufacturing Skills Certification System. The NIST Manufacturing Extension Partnership (MEP), MEP state centers, and the NAM Manufacturing Institute encourage manufacturers to use the credentialing system in job announcements and in job applicant screening. MEP state centers, the Manufacturing Institute, key workforce partners, and community colleges partner to ensure that their states have the appropriate policies and infrastructures in place for widespread use of the credentialing system to help manufacturers optimize workforce talent.

Department of Education (ED)

The Department of Education *National Career Clusters Framework* supports quality career and technical education programs through learning and comprehensive programs of study (POS). As one of 16 Career Clusters in the Framework, the manufacturing cluster guides development of programs of study in manufacturing that bridge secondary and postsecondary curricula. The manufacturing cluster creates individual student plans for a complete range of manufacturing career options. Career Clusters help students discover their interests and passions and empower them to choose an educational pathway that leads to success in high school, college, and career.

Department of Labor, Employment and Training Administration (ETA)

The Department of Labor's Employment and Training Administration supports the development of a skilled manufacturing workforce through the Registered Apprenticeship program, the Workforce Investment Act programs, and the manufacturing competency model. The Registered Apprenticeship program provides employment and a combination of on-the-job learning with a mentor, technical and theoretical instruction, and progress-driven wage increases. The program meets the needs of high-growth industries by keeping workers' skills updated in response to new technology. Today, there are over 3,000 apprenticeship programs and approximately 17,000 apprentices in manufacturing. The Workforce Investment Act (WIA) programs provide job training and employment services to job seekers and respond to the needs of employers for skilled workers. In the last program year, 19,000 participants who exited the WIA Adult and Dislocated Worker programs received training in occupations in the manufacturing sector. ETA released an updated manufacturing competency model (Appendix E) in spring 2010. Working with industry partners such as the National Association of Manufacturers, the National Council for Advanced Manufacturing, and the Society of Manufacturing Engineers, this employer-validated model outlines the skills necessary to pursue a successful career in the manufacturing industry. The model affords workers in manufacturing fields the ability to advance their training in a way that is consistent with industry demands.

National Science Foundation (NSF)

NSF supports graduate and undergraduate students through its *Graduate Research Fellowship Program* (GFRP) and through relevant competitively awarded research grants. The Advanced Technological Education (ATE) program supports community colleges partnering with industry and others to respond to industry needs for highly qualified manufacturing technicians (see main report text). NSF is creating the next generation of scientific and engineering manufacturing leaders through innovative partnership agreements to provide augmented educational experiences, such as the NSF-wide Grant Opportunities for Academic Liaison with Industry (GOALI), Partnerships for Innovation (PFI), and Accelerating Innovation Research (AIR). These programs provide hands-on educational opportunities for students and/or new entrepreneurs to learn about product development, product qualification, and scale-up of manufacturing.

Appendix E: The Advanced Manufacturing Competency Model of the Department of Labor

The competency model framework for Advanced Manufacturing (Figure 3) was developed through a collaborative effort involving the DOL Employment and Training Administration (ETA) and leading industry organizations, including the National Association of Manufacturers, the National Council for Advanced Manufacturing, and the Society of Manufacturing Engineers. To ensure that the model reflects the knowledge and skills needed by today's manufacturing workforce, ETA worked with its industry partners to update the original model. The updated model was completed in April 2010, and contains new information on Sustainable and Green Manufacturing, as well as updated key behaviors in several competency areas.

http://www.careeronestop.org/competencymodel/pyramid.aspx?hg=Y

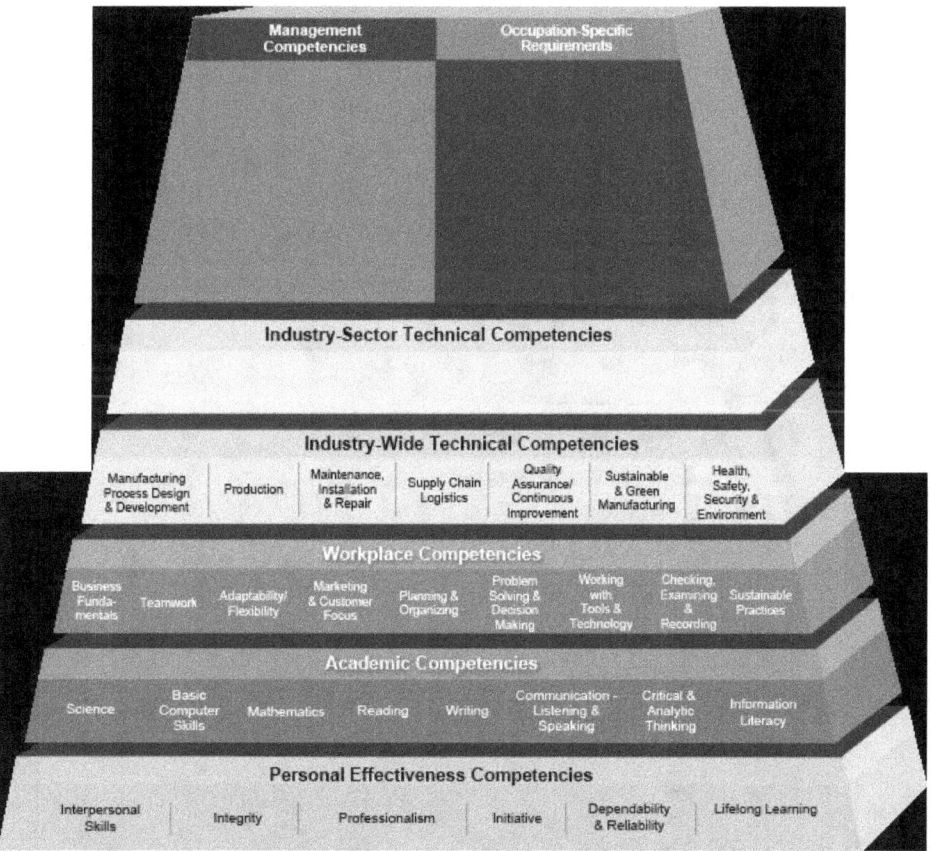

Figure 3. 2010 Advanced Manufacturing Competency Model

Appendix F: Examples of Federal Investments in Advanced Manufacturing R&D

Examples of Federal Investment in Advanced Manufacturing Processes

DOD DARPA	Antibody Technology Program
DOD DARPA	Open Manufacturing
DOD DARPA	Disruptive Manufacturing Technologies
DOD DARPA	HEALICS (Self-HEALing Mixed-Signal Integrated Circuits)
DOD DARPA	Tip-based Nanofabrication
DOD ManTech	Rapid Manufacturing of Aerospace Structures
DOD ManTech	Prosthetics and Orthotics Manufacturing Initiative
DOD ManTech	Joint Strike Fighter Producibility and Affordability Initiatives
DOE EERE	Innovative Manufacturing Initiative – Innovative Manufacturing Processes
EPA	Nanotechnology Research Program
EPA	Green Chemistry Awards
EPA	Energy Star
NASA	Next Generation Robotics
NASA	EBF3 Electron-Beam Freeform Process Fabrication Demonstration
NASA	Strong Cities, Strong Communities
NIST	Next-Generation Robotics and Automation
NIST	Integrated Smart Manufacturing Processes, Equipment, and Systems
NSF	Core Research Programs in Multiple Directorates
NSF	Engineering Research Center Program
NSF	Industry/University Cooperative Research Center Program
NSF	Grant Opportunities for Academic Liaison with Industry (GOALI) Program
USDA	Biomass Research and Development Initiative
USDA	Specialty Crop Research Initiative
USDA	McIntire-Stennis Formula Grants
USDA	Hatch Act Formula Funds, Evan-Allen Formula Funds, Smith-Lever Formula Funds, Extension Programs for 1890 Institutions Formula Funds
USDA	Bioenergy
USDA	Bioenergy for Advanced Biofuels
USDA	Biorefinery Assistance Program

APPENDIX F: EXAMPLES OF FEDERAL INVESTMENTS IN ADVANCED MANUFACTURING R&D

Examples of Federal Investment in Advanced Materials

DOD DARPA	Hybrid Multi-Material Rotor
DOD DARPA	Lightweight Ceramic Armor
DOD DARPA	Structural Logic
DOD ManTech & DPA Title III	Producibility of Large, Affordable Substrates for Advanced Imaging
DPA Title III	Carbon Nanotube Development & Production Enhancement
DOD ManTech	Advanced Body Armor Material Production Technologies
DOD ManTech	Affordable, Advanced Titanium Powder Processing
DOE ARPA-E	BEEST (Batteries for Electrical Energy Storage for Transportation)
DOE EERE	Innovative Manufacturing Initiative – Innovative Materials
NIST	Next-Generation Materials Measurements, Modeling, and Simulation
NSF	Advanced Materials Genome Initiative
NSF	Materials by Design
NSF	Materials Research Science and Engineering Centers

Examples of Federal Investment in Product Technology Platforms

DOD DARPA	Manufacturable Gradient Index Optics
DOD DARPA	IRIS (Integrity and Reliability of Integrated Circuits)
DOD DARPA	GRATE (Gratings of Regular Arrays and Trim Exposures)
DOD DARPA	Maskless Nanowriter
DOD ManTech	Advanced Electronics Packaging and Fabrication
DOD ManTech	Lithium-ion Battery Producibility Advancements
DOE ARPA-E	Direct Wafer Technology
DOE EERE	Innovative Manufacturing Initiative – Innovative Manufacturing Processes
NIST	Biomanufacturing
NIST	Nanomanufacturing
NIST	Advanced Semiconductor Electronics Manufacturing
NIST	Flexible Electronics Manufacturing
NSF	Nanomanufacturing
NSF	Bio-Economy
NSF	Cyber Physical Systems

Examples of Federal Investment in Data and Design Infrastructure

DOD DARPA	ADAPT
DOD DARPA	META
DOD DARPA	iFAB (instant Foundry, Adaptive through Bits)
DOD DARPA	FANG (Fast Adaptable Next generation Ground vehicle)
DOD DARPA	Diverse and Accessible Heterogeneous Integration
DOD ManTech	Connecting American Manufacturing Initiative
DOD ManTech	Advanced Supply Chain Modeling
DPA Title III	Heavy Forge Production Capacity Improvement
DHS	SECURE (System Efficacy through Commercialization Utilization Relevance and Evaluation)
DHS	FutureTECH
DHS	SAFETY Act (Support Anti-terrorism by Fostering Effective Technologies Act)
EPA	Lean Manufacturing
EPA	Green Engineering
EPA	Design for Environment
NIST	Sustainable Manufacturing
NSF	Engineering Research Centers
NSF	Industry/University Cooperative Research Centers
NSF	Innovation Corps
USDA	Quality and Utilization of Agricultural Products
USDA	Forest Products Utilization